HOW DO YOU KNOW YOU ARE RIGHT?

A Discovery

by

Andrew J. Galambos

REVIEWED BY

THOMAS SNOUSE

How Do You Know You Are Right?

FIRST EDITION
DECEMBER 2015

ISBN-13: 978-1519421722
ISBN-10: 1519421729

SCOT'S PRINT
LOMPOC, CALIFORNIA

How Do You Know You Are Right?

Table of Contents

How Do You Know You Are Right?

*What Do **You** Care What Other People Think?*
~ Richard Feynman

Preface

Feynman wrote, "It seems to me that scientists do think about [social] problems from time to time, but we don't put a full-time effort into them — the reasons being that we know we don't have any magic formula for solving social problems, that social problems are very much harder than scientific ones, and that we usually don't get anywhere when we do think about them."

"I believe that a scientist looking at nonscientific problems is just as dumb as the next guy — and when he talks about a nonscientific matter, he sounds as naive as anyone untrained in the matter."[1]

The reviewer admires Dr. Feynman in the extreme. This, however, is what I would tell him. The Laws of Nature are universal in our

---

[1] *What Do **You** Care What Other People Think?* Richard P. Feynman, W. W. Norton Company 1988 p 240.

cosmos. The universe is comprehensible.[2] Man is part of the Universe. Hence the Scientific Method is applicable to human behavior and it is a mistake to abandon a search for knowledge by labeling it 'nonscientific'.

T. Snouse

---

[2] This is the first postulate of the integrated super science of physics, biology and volition as suggested by Andrew J. Galambos.

The Scientific Method is
common sense crystallized.
~ Andrew J. Galambos

Right in Physics

How do you know you are right in physics?
It is actually quite simple in concept: if something is simultaneously true and valid it is right
on an absolute basis.

Please note; until we define terms, the
above statement is not very useful. Why? Because words often mean different things to different people. Imprecise or multiple definitions
lead to flawed communication which, in turn,
leads to flawed thinking.

This is no longer a problem in physics
thanks to Newton and others who realized that
unique definitions are an essential part of problem solving. Hence we have one and only one
meaning for terms such as mass, momentum
and acceleration. These concepts may be ex-
pressed in different units just as pi can be ex-
pressed numerically using different bases. Different units or bases, however, do not change
nor blur concepts. So, let's define terms. All

definitions are those used by Andrew J. Galambos unless otherwise noted.

True: that which is observable.

Observation is operationally defined and is the tie between the building blocks of nature and the thoughts and ideas of mankind. In other words, a scientist must document a result so others may test it. Physics is easy in many respects because inanimate matter is so very patient. One can run experiments for years. Other people, given the details of the experiment, can repeat it for corroboration. For one-time events, such as the collision of a comet with Jupiter, multiple observers can make simultaneous observations and then compare. Observation may or may not be easy, it may be expensive, but it is absolutely essential to the concept of rightness in physics. As Arthur Eddington said, "For the truth of the conclusions of physical science, observation is the Supreme Court of Appeal." Suffice to say, it is of mutual benefit to scientists to understand and resolve any differences in observation that may arise because of experimental error.

Galambos stresses that a truth is an item of fact. It is not an opinion. Truth consists of the rules of reality. Reality is the way the world and the universe are. Unfortunately, the terms *true* and *truth* are often used outside the scientific community to mislead or mystify. For the purposes of this discussion we'll stick to observation. We might note that the words *faith* and *belief* are the appropriate terms to use when observation is lacking.

Valid: conforming to the rules of logic.

The rules of logic are well understood and there is no need for a lengthy discussion of them. In practice one checks one's logic by running through a checklist of logical errors. Is there a misidentification of cause and effect? Have steps one and two of the Scientific Method been reversed? Has an assumption been made which is not supported by observation? Is the reasoning circular? And, sadly, could someone have fiddled the data?

It is important to note that the concept of validity is not independent of truth, it is an epistemological derivative of truth. That is, logic enables access to items that cannot be directly or readily observed. Put simply, logic is mental machinery, tested by time, that outputs truths (observables) when truths are input into a mental process. Logic and observation have produced useful knowledge about entities which we will likely never observe directly, such as electrons and the interior of stars.

We know that some logical and mathematical thought processes, such as those used in string theory, are so convoluted, esoteric and downright difficult that one might ask how we know they are valid? We will know they are valid when their predictions agree with observation one hundred per cent of the time. Thus *true* refers to starting observations of reality, operationally defined, while *valid* refers to the thought processes and the truth output therefrom. *Truth* is direct observation; *valid* is truth predicted and observed after mental effort.

Note: bear in mind that thought processes and results must also be consistent with the basic postulates of each scientific endeavor.

Absolute: independent of arbitrary standards of measurement.

What does this mean? In practice, it refers to a physical quantity that has the same value no matter how competent experimenters choose to measure it. As far as we know today, the speed of light in a vacuum is an absolute. Planck's constant is another. Pi is a third. In one sense, physics can be described as a search for absolutes in a universe of seeming chaos and change. Absolutes have been a key to progress in the physical sciences and a scientist is lucky indeed to find one.

So there you have it: in physics something is **right** on an **absolute** basis if it is both **true** and **valid**.

As long as we are happily defining terms, let's try three more; theory, corroboration and hypothesis.

Hypothesis: a guess proposed as a theoretical explanation for the occurrence of some specified group of phenomena.

Corroboration: repeated agreement of experiment with hypothesis with no exceptions. (In other words, aside from experimental error, if the results are not 100% as the hypothesis predicts then you go back to the laboratory.)

Theory: a multiply corroborated hypothesis.

This draws a useful distinction. In the sloppy world of the media and social interaction you find *theory* and *hypothesis* used interchangeably. Half-baked ideas are accorded the status of theories whilst ideas which have stood the test of time are dismissed as old fashioned. This, of course, gave rise to Santayana's famous quote that those who don't know history are doomed to repeat it.

One more:

Law of Nature: an honor bestowed on a theory so multiply proven over time that it is no longer a topic of serious debate. Newton's Law of Gravity is a fine example of a Law of Nature.

"Oops," you may be thinking, "didn't Einstein's Theory of General Relativity disprove Newton's Law of Gravity?"

"Sorry," is the reply. "Check with any physicist and you'll be told that Einstein's equations reduce to Newton's equations at low velocities. What Einstein did was to extend our knowledge. He showed the limits of Newton's equations and reinforced their rightness within those limits."

I might add that science is never static. As a famous physicist once said, the process of discovery is like the layers of an onion; once you dissect one layer you find there's a deeper level to explore. Hence, physicists are now working on a larger hypothesis which will encompass gravitation and quantum electrodynamics — but still reduce to the same equations within the appropriate domains.

Perhaps this is a good spot to deal with some predictable quibbles and misconceptions. Probably the most popular quibble comes from people who are habitually vague. "I don't care for your definitions. I'm afraid I could be misled because they differ from the definitions I use."

Many times the background which causes this concern is due to the well known evil that can be done by labeling in the social arena. If you define Jews, as some Islamic fanatics do, as people who drink the blood of Muslim babies — well then you can expect trouble.

How Do You Know You Are Right?

The above is an offensive lie for political purposes, not a definition. Yet it has a utility beyond its abhorrent persistence. The utility is this: the fanatic has communicated his fantasy and thus you know his point of view and intentions. By no means do you need to adopt his definitions nor — I might add — does he wish to adopt yours. But if you offer your definition to a fanatic — should you care to — then at least there will be communication. Not agreement, mind you, but the respective points of view should be clear.

The same is true for the above definitions. You need not adopt them unless you find them superior to prior usage. Take your time. Meanwhile, for the purposes of communication, you should, hopefully, find this exposition semantically precise.

In other words, Galambos' definitions have the aim and the virtue of clearly separating that which is defined from everything else in the Universe. They are not designed to mislead nor to mislabel. They are necessary parts of a new intellectual tool.

There is nothing more interesting or more important in science than the observations that we cannot explain.
~ Thomas Gold

Property as an Integrating Factor

Physics is easy compared to biology or social science. Probably an order of magnitude[3] easier than biology and two orders easier than social science. We've already mentioned one reason physics is easy: inanimate matter, by and large, is readily manipulated. But there's another, more important, reason: Isaac Newton's breakthrough discovery of previously unseen relationships between the separate areas of physics. Physics, in his day, was called natural philosophy and natural philosophers studied mechanics and the precursors of disciplines such as thermodynamics as if they were rather different animals. After Newton integrated the physical sciences with the publication of his magnificent *Principia Mathematica,* scientists began to see the many connections between dis-

---

3 An order of magnitude is ten times bigger or smaller or harder, etc.

ciplines and to use the new awareness to create a veritable explosion of new knowledge.

One outstanding example deserves mention. Edmund Halley (pronounced Hawley) used Newton's equations and the astronomical observations of John Flamsteed to predict the return of the comet that now bears his name. Prior to Halley and Newton, comets were objects of dread — harbingers of disaster. Since disasters were fairly plentiful in those days it wasn't too difficult to ascribe a war or the death of a king to the comet *du jour*. Added to that was a belief that a comet's tail was highly poisonous.

When Halley's prediction of the comet's return came true, it then became apparent that a comet was simply another heavenly body in orbit around the sun. It took decades, of course, for that fact to be accepted by the public. But in the long run it certainly boosted the stock of science and scientists. **We would not have the technology of today had it not been for Newton and his intellectual heirs.**

This leads us to an interesting comparison between the physical sciences and the social sciences of today. In physics, believe it or not, we have more questions than answers. This is because our sphere of knowledge has been expanding nicely. Thus the surface area of the sphere is larger today than it was a few years ago. The surface area, the boundary of knowledge, if you will, is where the questions are.

In the social arena what do we know for sure? We have 'experts' who claim to have most of the answers but few worthwhile questions. We do know we have too many societal diseases such as war, theft and slavery[4]. How are they related? Is there an integrating factor? Can greater knowledge lead mankind to a future in which there is no possibility of mass murder as practiced in past centuries? (This century too, more's the pity.)

It is tempting to build up suspense for a few more pages, but since this is a physics book, here it is: **property**, properly defined, is the integrating factor in the social domain. The definition of property will come in a moment, but this is a good place to note that energy, properly defined to include all its various forms, is the integrating factor in the physical sciences.

---

[4] Slavery, properly defined, is far more common than you might think.

17

If something moves, there is kinetic energy to be measured. If something doesn't move it usually has thermal and potential energy. Energy is released or absorbed in chemical reactions and the knowledge of those energy transactions gives competent chemists the ability to build custom compounds with the ease of a child playing with Lego. Electrical energy alone has fueled at least two revolutions in our standard of living.

One could go on at great length about this subject and how, interestingly, most major scientists showed a strong childhood fascination with energy and its many manifestations. Follow this up elsewhere if you like, but now it is time to give Galambos' definition of property.

Property: individual man's life and all non-procreative derivatives of his life.

As you may note, this is not the definition you will find in your *Funk & Wagnalls*. It is at once simple, inclusive and very useful. We shall see some delightful examples presently.

Just as there are different kinds of energy, there are different kinds of property. The first, which Galambos calls **primordial**

property, is your life. The next kind (**primary property**) consists of your thoughts and ideas. (An idea is an articulated thought.) Ideas lead to actions which in turn often produce tangible forms of property called **secondary property**.

Let's explore the hierarchy. Clearly, you must first be alive to own anything else. So, owning your life is a necessary condition for you to generate and subsequently own and enjoy the other two kinds of property. Primary property (thoughts and ideas) ranks ahead of secondary property because secondary property cannot be produced without thought. One way to look at it is: which would you rather lose, the accumulated fruits of your labors or the knowledge of how to produce those fruits?

One reason this definition of property is useful is the simplification it offers in analyzing social transactions. For example, all disputes are now property disputes. Once all the relevant properties and owners have been identified in a dispute, it then becomes a straightforward procedure to reach a just settlement. All exchanges are property transfers. More about this later.

The concept of ownership is inextricably associated with the concept of property. The person who creates something, whether it is an idea, a sculpture or a jar of white lightening, is the property owner. Please note, property can be positive or negative. Should you have the misfortune to be at fault in an auto accident, you now own the responsibility of reversing the damage done to the other party. This can be a major example of negative property. Thus prudent people pay for insurance to avoid a possible financial disaster. As a property exchange, one sees that the insurance company balances the unknown probability of a claim (negative property) in return for a known amount of positive property, the premium. Property can be jointly owned by more than one individual. Individuals holding stock in a modern corporation would be an example of multiple ownership.

Are children property? No, they are procreative derivatives of the lives of two individuals. Children are the owners of their own primordial property and for another individual to claim to own them would be slavery. However, parents do have a major ownership position with regard to their children. Parents own the responsibility to support, educate, discipline and assist their child until it is competent to survive on its own. Obviously, this criterion varies widely from family to family and culture to culture. One way to justify this responsibility is to realize that you owe a debt to your parents for the support they provided you as a child. It can be a satisfying experience to discharge such a debt.

Is land (aka real estate) property? No. Land is *not* property because it is not derivative of any person's life. However, the *use* of land (properly delineated from an initial claim and later purchased) *is* derivative and *is* property. Improvements such as a house, a sewer line, a driveway are property and add to the perceived value associated with use of the land.

There are three ways property can be transferred. The most important way is with the consent of the owner. One can get this consent by offering an exchange. One could offer a good, a service, currency or even money.[5] A gift is also a satisfying property exchange. The recipient gets something of value, the giver gets a primary property addition to his or her happiness.

---

[5] Currency and money are not the same thing. The former is an IOU, the latter has intrinsic value.

How Do You Know You Are Right?

A second form of property transfer, regrettably all too frequent, is theft or involuntary transfer. Since stolen property is no longer under the control of its owner, it is no longer property, it is plunder. The owner is entitled to recovery and, should that happen, it becomes property again.

A third form of property transfer is loss or discard. Loss is often due to negligence. In fact, a form of double negligence is common. If a tangible item carries the owner's name and contact information it may well be recovered. If not, it is then in a state of non-ownership and may be claimed as property by an honest finder after a reasonable effort to find an owner fails.

Discarding unwanted property can be right or not right. Paying for garbage pickup is right, discarding as little as an empty soda cup on someone else's real estate is not right.

It is the tension between creativity and skepticism that has produced the stunning and unexpected findings of science.
~ Carl Sagan

Right in Volition

So . . . what does property (properly defined) have to do with physics and human decisions? A physicist uses the intellectual and mechanical tools of physics to track the flow of energy in order to solve problems. Similarly, a scientist in the field of human volition can follow a similar bookkeeping procedure to track property — in order to solve problems.

To recapitulate, Galambos writes, "The criterion of rightness is simply the totality of truth and validity. That is absolutely right in physics which satisfies simultaneously the truthful nature of all input propositions and all conclusions that are observationally corroborable. At the same time, all thought processes employed in arriving at these conclusions conform to the tests for validity. These are determinable in terms of thought processes which consistently, reliably, and with zero failures are capable of producing successful conclusions. Which means that they are capable of being corroborated by observation as to their truth content. If all the premises and all the conclusions are true and if the thought processes employed are valid, then that which has been arrived at as a conclusion is absolutely right."

Can this intellectual tool be expanded to encompass the enormous set of possibilities in the social domain generated by the vagaries of sentient decisions? Can it encompass irrationality as well as rationality? Of course it **must** because irrationality is clearly an observable! But first we need to carefully delineate concepts with precision definitions. Once again, a useful definition separates that which is described from all other items in the universe.

Good: an individual's subjective evaluation of a (volitional) preference.

Bad: an individual's subjective evaluation of a (volitional) dispreference.

Happiness: **the sum of an individual's goods minus the sum of their *bads* to date.**[6]

Coercion: attempted, intentional interference with the property of another.

There are two methods commonly used to induce property transfer or volitional compliance; force and fraud. **Force** includes physical harm or the threat thereof including imprisonment. **Fraud** is intellectual deception on any level. You voluntarily give up your property for a promise of a benefit that either isn't true or isn't delivered as promised.

Absolute Good: a subjective good to at least one person, which is not imposed on another volitional being.

'Absolute' is used to identify the concept that coercion is entirely absent regardless of the identity of the individual specifying the good. Please note that the identification of what is good clearly varies from individual to individual as well as from time to time for the same individual or individuals. Given its subjectivity,

---

[6] The Law of Logarithmic Stimulation reminds us that current goods and bads are weighted more heavily in this evaluation than those in the past.

we realize that ice cream might be a good to most people — but not to people who are lactose intolerant. An absolute good may be as trivial as ice cream or as important as liberty.

Moral: coercion is absent.

Galambos suggests that morality is the name of the subject that deals with absolute goods — the totality of which are linked to one another through a single concept, the fact that coercion is absent. Note: this definition is not derived from theology or case law.[7]

Add this to the standard of rightness in physics and you get the absolute standard of rightness with respect to human volition.

Something is **right** on an **absolute** basis if it is simultaneously **true**, **valid** and **moral**.

Please pause for a moment and think about this statement. Don't let it slip by without question. It may be the most significant discovery made in three centuries. Repeating for emphasis:

---

[7] The reviewer finds it ironic that the phrase "Due process of law" requires a sixty-seven-page discussion in a legal dictionary.

How Do You Know You Are Right?

Something is **right** on an **absolute** basis if it is simultaneously **true**, **valid** and **moral**.

Something is wrong at this point in your life if you are not suffering from some degree of cognitive dissonance. Why? Because we have all been immersed since birth in a milieu which, to put it bluntly, is selectively immoral. Just as a cannibal sees nothing wrong with a ladyfinger lunch, we tolerate plunder as a fact of life. Some of us express our cynicism through statements such as the following: "How else could it be? That's just the way things are. Get real. You can't sue city hall. There's nothing new under the sun. If only people would share. Love is the answer. You can't save the world. I'm all right, Jack. If it's so great, somebody would have thought of it before. It can't be a discovery, I always thought that way. Besides, you can't change human nature."

Only one of the above statements is right and we'll deal with it in the next chapter. All of the statements are annoying impediments to progress. Inventors and innovators are, unfortunately, very familiar with negativism and rejection. They know it takes time for people to understand and use new ideas and technologies. And, the bigger the discovery, the longer it often takes. Newton's ideas, we are told, did not make it into the classroom for decades after his discoveries. Mendel's genetic discoveries were ignored for twenty years and only became known because of the integrity of researchers Hugo de Vries and Carl Correns who independently made the same discoveries.

Galambos' ideas have been extant for decades but have yet to reach the right intellectual market. This short discussion is aimed at people who are curious, intellectually honest and irrevocably committed to the use of the Scientific Method. Galambos' standard of rightness will greatly augment your ability to think and to solve volitional problems, but it will not change human nature.

To mistrust science and deny the validity of the scientific method is to resign your job as a human. You'd better go look for work as a plant or a wild animal.
~ P.J. O'Rourke

Happiness and Human Nature

'Human nature' is a wonderfully vague, all-purpose term. It belongs to a group of glib phrases such as 'common good,' 'social contract,' 'your civic responsibility,' that we use routinely and think we understand. Yet all of them are major topics of philosophic debate without sound foundation. Thomas Paine said he could not understand the concept of the 'common good' unless it was simply the sum of all the personal goods as measured by each of the individuals in a community. For an individual or individuals to order other individuals to sacrifice lives and treasure for a 'common good' is coercive, hence not right. 'Common good' as a concept certainly sells — perhaps because we wish to believe we shall be the beneficiaries of the common good rather than the ones who sacrifice.

How Do You Know You Are Right?

Please note; there are always many ways to achieve a desirable goal. The challenge is to find non-coercive methods to achieve that goal. The use of coercion is certainly a simple and direct way of doing such things as national defense. This method has been in use since we left the caves and while it works to a limited extent, it has also produced tyrant after tyrant who have claimed the right to kill and murder in the name of the common good. Sometimes tyrants kill their neighbors and sometimes kill their citizens. Rather tarnishes the concept of common good, doesn't it? A better concept of the common good would be total protection of every individual's property — life, intellectual property and tangible property. This concept is, of course, called freedom. Currently, we are far from it.

'Social contract' is a similar buzzword without validity. Contracts, to be binding, must be voluntary. Lysander Spooner once asked, "How can society make a binding contract for an unborn individual?"[8] 'Civic responsibility' is a similar try at a pseudo-contract and is, in fact, a

---

[8] No Treason No. VI: The Constitution of No Authority, 1870, Lysander Spooner

mild, well-meaning form of fraud, hence not right.

Human nature is, to a degree, measurable in the aggregate. We measure crime rates, savings rates, percentage of households with children and/or pets, educational degrees — the list is long and boring. Statistics tend to confirm what we already know: some people are criminals, while some save money, have children and/or pets and have gone to school. Not entirely helpful if we wish to predict the behavior of individuals.

What is needed for understanding is a deeper and simpler look at the universals of human behavior. Galambos points out that all sciences start with postulates. Some call them axioms but he prefers the term postulates. What, he says, is a postulate? A postulate is a proposition with a truth content which you did not derive from any earlier set of propositions by logical reasoning, but whose validity you accept as an input into the subject you are developing. It has to be true — which is not a logical but an observational concept. To be a postulate, the proposition must have a truth content which is without exception. Then you may accept it as being valid without having derived it from anything else.[9]

Postulate number one: All volitional beings live to pursue happiness.

Postulate number two: All concepts of happiness pursued through moral action are equally valid.

Galambos states that he spent eight years testing and discarding propositions as candidates for postulates. The winning postulates have a deceptive simplicity which is certain to invite criticism. Before we deal with common criticisms we should remind ourselves, á la Occam's Razor, that simplicity and elegance are often hallmarks of a successful theory in physics. The enormous multiplicity of variables in volitional science is not sufficient reason to abandon criteria which have served science well in the past.

Galambos makes a distinction between happiness and the *pursuit* of happiness. He credits Thomas Paine for recognizing this im-

---

[9] Sic Itur ad Astra, Andrew J. Galambos, The Universal Scientific Publications Co., Inc., ISBN 0—88078—002—9, p 95

portant distinction. Happiness, as we all know, comes in all sizes, shapes and tastes. What makes one individual happy can be disgusting to another. What makes the same individual happy one day can leave him or her cold the next. This is often a preference based on emotion rather than reason. Isn't that part of human nature?

But the pursuit of happiness is different. If happiness is the goal (not always achieved nor rationally pursued), then the pursuit of happiness is the direction and the means taken to reach that goal. What about multiple goals? Happiness goals can range from the trivial to the sublime. We set priorities, don't we? Geniuses may spend twelve or more hours a day working on an equation, a symphony or a work of art but still they must brush their teeth, go to the bathroom and have a bite to eat. In other words, their pursuit isn't a straight line towards their most important source of happiness, it is often detoured by the realities of life such as having a place to live and paying taxes.

Most of the objections to the first postulate are related to our failure to understand people whose motivations, thinking and personal situations depart from the norm. What about the individual who commits suicide? Is he seeking happiness? One can easily empathize with someone who has a very painful fatal disease, such as certain cancers. At some point a sufferer may decide the pain of living is worse than the unknown of death. Understandable. What about healthy, successful, intelligent individuals who commit suicide? Surely they are not seeking happiness? Obviously, we have a multiplicity of situations which lead to an individual's decision to put an end to it all (some of which are often portrayed on daytime TV).

Two case histories may help us empathize. In the last century, it was fairly common for high achievers to maximize their productive time by taking stimulants in the morning and barbiturates at bedtime. They were happy to get more done as a result of chemical empowerment. The catch was: alcohol and barbiturates taken at bedtime can be fatal. More than one show business celebrity and more than one university professor have met their maker inadvertently. Please note, their pursuit of a minor happiness, namely a good night's sleep, caused the loss of a major happiness: their life.

A second case history is a situation labeled as suicide which may not be suicide at all.

How Do You Know You Are Right?

A recent (2012) large (over 45,000 data points) epidemiological study of antibodies to Toxoplasma gondii in the blood of newborns in Denmark was associated with a significantly higher risk of suicide by those mothers whose children had these antibodies. This protozoan can infect humans by several routes, including unwashed vegetables, raw meat and inadvertent exposure to the feces of infected cats.

The toxoplasmosis parasite has a life cycle divided between cats, rats and, presumably, mice. It works like this: a rodent is infected by the parasite's egg shed by a cat. The cat, by the way, is seldom harmed by the parasite. But the rodent is. The parasite rewires its brain with the result that the rodent loses its natural fear of predators. Presto, instant cat food. The parasite then completes its development in the cat's intestines in order to lay eggs. If a human is infected it is probably a dead end for the parasite. There is reason to believe it is also a dead end for a very small group of people who, like the mice, lose their fear of death. Thus, what appears to be a suicide may not be one, just as some fatal auto accidents may be suicides in disguise. This is a very sad finding because, if true, medical science might have been able to prevent some otherwise inexplicable deaths.[10] In addition, it's sad because Galambos was extremely fond of cats. I doubt he would have given up his pet even if it carried a knife.

A more important question than suicide is this: what if an individual's concept of happiness is sociopathic? Sociopathic behavior is practiced by roughly two percent of the population. As suggested by Paul Rosenberg,[11] hallmarks of this behavioral problem are lack of empathy for others and a defective or non existent conscience. A well-known example comes from those jailed for major crimes. Psychological testing has shown at least one common denominator possessed by societal rejects: they very often blame the victim. This lack of empathy enables them to escape the blame (in their own mind) for the damage they do to others. "Society made me do it." Or, "She shouldn't have walked home alone from the bar."

---

[10] Pedersen, M.G., Mortensen, P.B., Norgaard-Pedersen, B. & Postolache, T.T. Toxoplasma Gondii Infection and Self-directed Violence in Mothers, *Archives of General Psychiatry,* DOI: 10.1001/archgenpsychiatry. 2012.668

[11] freemansperspective.com

How Do You Know You Are Right?

Not all sociopaths are in prison. Some are better able to calculate the risk/reward aspects of their efforts to plunder and control the property of others. They learn to counterfeit normal behavior in order to get what they want, be it sex, political power or wealth. Some sociopaths are minor leaguers who are content with using controlling behavior toward their family and spouse. Some sociopaths progress (if you can call it progress) to psychotic behavior — Hitler, Stalin, Mao Tse Tung, Pol Pot, Genghis Khan — the list is too damned long of nut cases who have used the mechanisms of the state to murder millions. Clearly a sharp line must be drawn which excludes harmful behavior toward others.

Nevertheless, with reference to the first postulate, it is important to remember it is the individual who defines his or her concept of happiness. It isn't you, your next door neighbor or the president of the Rotary Club, it is the individual. All three of you plus the Chief Justice of The Supreme Court may find the individual's concept of happiness trivial or deplorable, but remember: the concept is the individual's property and he or she is stuck with it. Note that an individual's ideas are but dreams until acted upon.

Actions can and most often do involve others. This brings us to the second postulate: all concepts of happiness pursued through moral action are equally valid. Put simply, one person's happiness goal is not to be judged as long as the pursuit of it does no harm to any other individual's property including trespass. As long as something is moral, there is no *preferred* moral frame of reference. Galambos suggested that this postulate is analogous to Einstein's statement to the effect that all frames of reference are equally valid for the purpose of formulating the laws of nature. He calls it the *democratic postulate* and the first postulate the *happiness postulate*.

Please note; the second postulate applies to the methods or actions taken in the pursuit of happiness. You will find it to be a revolutionary departure from current societal thinking in that **there are no gray areas when it comes to morality.** You either own (or have authority from the owner to control) the property under discussion or you don't. And . . . your actions have either harmed the property of others or they have not. "Too simple!" you may say as your initial reaction. "I certainly don't like it when my property is taken without my consent, but I usually get something in return." Yes, you do: schools, roads, fire and police protection, welfare, wars, inflation and sometimes tyranny. This statement may strike most people as unfair in the extreme but it is verifiable.

Some individuals will take issue with the idea that there are no gray areas permitted by Galambos' definition of morality. They can, of course, define morality in any way they like. Let us consider the implications of a relative morality — one which permits grey areas. Instead of asking, "Whose property is it?" the question becomes, "Who can seize property with impunity, how much can they demand and how can this behavior be circumscribed?" Isn't this a question our species has faced since we learned to walk on two legs?

Robert Ardrey wrote, "Among primate societies establishment of dominance is universal and may result not only in male hierarchies but, as in the Japanese monkey, female hierarchies as well."[12] He was talking about social groups of primates including humans. Dominance is usu-

---

[12] The Hunting Hypothesis, p.86, Bantam Books, 1977

ally established by fighting, tests of strength or posturing. Variations abound with regard to the rules of contest but the observable result is pretty much the same: the winner makes the rules and get first choice of goodies, whether food, sleeping spot or sexual favors. Notice that this idea seems to still be the case today for humans, albeit employing far more sophisticated forms of competition and offering somewhat different rewards. The tools political humans use to achieve dominance and which enable them to make the rules for the rest of us are force and fraud. It seems we haven't truly outgrown our heritage from the African savanna.

On the savanna, dominance is not necessarily a bad thing for a group because when the trauma of establishing a pecking order is over, individuals can get on with their lives in relative peace. Constant contests for dominance would weaken the group and adversely affect the probability of the group's survival. Being a part of a well run group can benefit an individual's survival but the individual does need to ask himself how much he is willing to pay for this benefit. Please note, there is a psychological cost one pays for subservience. Perhaps, with the aid of science, we can abandon these ancient behaviors inherited from our animal ancestors and cease paying tribute to bullies. Once upon a time, bullies had value because they took the lead in dealing with predators and hostile neighbors. These are not necessarily our major threats today. Can bullies deal with an Ebola epidemic? And, just perhaps, can we rise above our instincts? Today we need brains, not bullies.

We have some useful rules of thumb in the social arena. Let's see how they connect with the postulates. Let's first take an easy one; Gresham's Law. This 'law' states that, in essence, bad money drives good money out of circulation if they exchange for the same price. Please note, it takes legal tender laws to force an equivalence that violates common sense. It is observable that Gresham's law has, over the long run, operated without fail in China, the United States, Europe, Zimbabwe and the Roman Empire. The postulates tell us why. As an example, in the USA when silver coins were replaced with copper-clad coins of the same denomination, individuals quickly took the silver coins out of circulation. One can still obtain a silver coin but it will cost you, as of this writing, thirty clad copper quarters to get a 90% silver half-dollar. People, in pursuit of happiness, are disinclined to give money away.

How about the law of supply and demand? Aside from special cases such as monopoly[13] and monopsony, it is observable that prices are inexorably tied to supply. Prices are bid up when an item is in short supply and prices usually drop when there's a glut. As a result of this natural mechanism, producers and consumers make adjustments. Producers see opportunity and ramp up production in the event of shortage. Retailers and wholesalers seek alternate suppliers whose prices, including transportation costs, may be lower. Consumers buy less or none at the higher prices and they, too, find substitute products. In other words, the pursuit of happiness impels both buyers and sellers to seek a middle price acceptable to both. In a free market the resultant prices and price trends are generated by billions of decisions made by individuals. This is true economic democracy and it is worth noting that it is al-

---

[13] Aside from single person monopolies, such as artists, many if not most monopolies are created by law. Professional licensure and college accreditation are two mechanisms which discourage competition.

most entirely free of problems such as racial or wealth based discrimination.

Here's an unfortunate fact: many political executives and legislators absolutely flunk Economics 101. They keep passing laws which attempt to repeal the natural law of supply and demand. They do so with the best of intentions, perhaps, but they may as well pass a law repealing the law of gravity. They may or may not be ignorant, but their propaganda succeeds because the public, in general, is even more ignorant. Bluntly, a plurality of the public is deceived into thinking their leaders can give them something for nothing. Not only is that thinking at odds with the law of supply and demand, it violates the laws of thermodynamics. Note that laws which interfere with normal market functioning will generate black markets and criminals. Would you rather purchase beer from Budweiser or Al Capone?

Another rule of thumb derivative from the postulates is the so-called Law of Unintended Consequences. When a bureaucrat uses coercion to advance a goal, the result is often opposite to that which is intended.

How Do You Know You Are Right?

In summary these postulates do not say that all concepts of happiness are equally important nor do they suggest that the happiness goal is always achievable. The postulates explain the failure of coercion, in the long run, to effect positive changes in social interactions.

New ideas in science are not right just because they are new. Nor are old ideas wrong just because they are old. A critical attitude is clearly required of every seeker of truth.
~ Thomas Gold

New Ideas

The bigger an idea, the longer it takes the world to digest it. After vicious intellectual debate a new generation delivers the ultimate accolade and insult, "Everybody knows that." Skepticism is certainly healthy but it wears hard on innovators. Remember Darwin?

The market for new ideas is exceeding small — unless any greedy fool can see their immediate value in cash or political power. This fact of life does not speak well of human beings and it is a negative factor when it comes to survival of the species.

Two examples from the world of medicine come to mind. Ignatz Semmelweis, a Hungarian doctor, was appalled by the number of new mothers who died from childbed fever in hospital clinics. At that time there were many hypotheses as to the cause. Some beliefs were patently silly such as night air, swamp gas and the like. Semmelweis noted that a woman had a better chance of avoiding childbed fever if she used a midwife rather than a physician. What was the difference between a birth at home and a birth at a maternity ward? He came to the conclusion that the evidence indicted the doctors. Somehow they were transferring contagion from one patient to another.

In that era, there was a touch of vanity exhibited by doctors. They wore the same clothing for an extended period of time and wiped their hands on their clothing. It was a form of advertising; the more blood and pus you had smeared on your apron, the larger your number of patients and, obviously, the better your reputation as a physician.

Semmelweis instituted a simple methodology in his maternity ward. He insisted that the doctors and nurses wash their hands before seeing a patient. He made the procedure stick and the death rate from childbed fever in his clinic dropped nearly to zero. Was he hailed as a hero? Of course not, his discovery was viewed by his peers as an affront! Why should he have a lower death rate when theirs was still high? Was he trying to blame the doctors? Yes! How dare he!

How Do You Know You Are Right?

This story has a mixed ending. Thanks to Semmelweis millions of mothers now live through childbirth. Thanks to critics and closed minded physicians, Semmelweis was hounded to a miserable early death in an institution.[14]

He had one ally. Joseph Lister, an English surgeon, had independently come to similar conclusions. They were in communication and it had to be gratifying for Semmelweis to know that his right ideas would live — even if he didn't.

Lister, too, had a belly full of opposition, disbelief and criticism. As a historical sidelight, he voyaged to America in 1876 to give a two hour exposition of his methods at the Centennial Fair held in Philadelphia. Although Lister's amazing results from the previous eleven years had silenced most of the critics in Europe, his methodology met with considerable skepticism from the American physicians. A quote from Dr. Samuel Gross, the president of the Medical Congress at that time and perhaps America's most famous surgeon, epitomized the prevalent skepticism. "Little, if any faith, is placed by any enlightened or experienced surgeon on this side

---

14 Wikipedia

of the Atlantic in the so-called carbolic acid treatment of Professor Lister."[15] The punch line to this anecdote; one of the physicians who attended Lister's lecture was later to attend James A. Garfield after he was shot by an assassin. Lister's antiseptic procedures were not used when the assassin's bullet was removed. President Garfield died of sepsis. But it was all very proper, presumably the president's physicians were properly licensed.

It may be that instinctive opposition to new ideas is built into the human genome. It may be that the rare individuals who exhibit strong intellectual curiosity and intellectual honesty are freaks of nature. These are the people, however, who have consistently led the rest of us out of the cave and into lives of comfort and health.

Often a new idea is criticized as being contrary to common sense. This is a pretty good bet, most of the time, given the miserable success rate of most ideas. Yet this argument can be overused. It is no substitute for genuine thought. First of all, what *is* 'common sense'? A common dictionary definition is "ordinary good

---

15 Destiny of the Republic, Candace Millard, Doubleday, 2011

sense or sound practical judgement." Somewhat circular and uninformative, wouldn't you say? Earlier in this narrative, Galambos was quoted as saying, "The Scientific Method is common sense crystallized." This is much more specific. Yet there is an aspect of common sense that needs discussion. It is this: common sense is largely based on the macroscopic world in which we live. We perceive that world through our five senses. This directly shapes our thought processes. When the scale of a phenomenon is very small or very large, common sense is no longer a trustworthy guide. Examples abound. Einstein's Theory of Relativity violates any-one's macroscopic world view yet GPS systems or particle accelerators cannot function properly unless they are designed using his equations. Experiments in the quantum domain which demonstrate the phenomenon called 'entangle-ment' definitely shred one's common sense.

And so, finally, we have an explanation for the opposition to Semmelweis, Pasteur and Lis-ter. People accustomed to using common sense had a difficult time accepting the hypothesis that there were living entities too small to be seen. Further, how could such small creatures harm people? In 1876 it was easier to believe in angels than to accept the existence of micro-scopic life — **even though the evidence was clearly there,** available to all who could see[16].

You might think we have wandered off top-ic. What do intellectual apathy and intellectual conservatism have to do with the concept of rightness? Just this: prior standards of rightness have been based on authority or legality. If you believe something is right because your profes-sor, your lawyer, your congressman or your pas-tor has passionately told you so — you then re-act badly to ideas which challenge conventional wisdom. On the other hand, if you realize that we need to improve our way of solving social problems, then you should investigate new ideas no matter how radical they seem at first glance.

When but one person is right it follows, as day follows night, that the rest of us are either ignorant or wrong. Isn't it easier for us to be in the majority and to dismiss or to ignore the ob-vious? Over millennia our species developed a herd mentality that had positive survival value at the time. The instinct for sticking with the

---

16 "You look but you do not see." Sherlock Holmes to Dr. Watson

herd may not be as valuable to the human race as it once was. For example, how much evil has been done by ordinary people who were 'just following orders'? We often 'go along to get along' even though we know our behavior may harm others. Douglas French pointed out that Edmund Burke's maxim: "The only necessary for the triumph of evil is for good men to do nothing" needs revision. He would add that evil is also served by good men and women who blindly follow orders. Sociopaths manipulate the herd just as a dairy farmer herds cows. Individuals **must** think for themselves else, unfortunately, they will have too much in common with cows.[17]

---

17 Aside from being milked twice a day, cows eventually get butchered.

History is a better guide than good intentions.
~ Jeane Kirkpatrick

Capital, Subset of Property

Were you to run a worldwide opinion poll on the two words "capitalism" and "profit" the result would likely be a landslide of negative votes. Put simply, it is to the advantage of people who believe in collectivist ideals to demonize present-day capitalists, profit and capitalism itself. Starting in kindergarten we are exhorted to 'share.' Later in school we are bombarded with examples of capitalistic greed and exploitation of employees. The media, short-termers all, chime in with biased reportage. There is a dreadful imbalance in this collectivist presentation since it omits or minimizes the fact that most of the necessities of life — food, gasoline, beer, chocolate, etc. — are produced by individuals and businesses seeking profit. The collectivists praise state control of the means of production despite the universal failure, in the long term, of those economies largely or completely run by bureaucrats. Capitalists, as defined by collectivists, are greedy, voracious bastards. Public officials, with their incessant demands for higher taxes, are not? This inaccurate characterization of the facts is widely accepted and is as close to pure propaganda as one is likely to get.

Collectivists are masters of the appeal to emotions. Emotions, for example, envy, are driving forces which 'legitimize' collectivism. As more and more people rely on emotion rather than rationality, propagandists find it easier to sell bad ideas such as the association of greed with capitalism and altruism with collectivism.

In reality the collectivist mantra, 'From each according to ability, to each according to need' is a blatant appeal to greed. Any citizen with the reasoning power of a toaster can readily glimpse the benefit of getting something for nothing. 'Something for nothing' is an excellent description of greed and a violation of the Second Law of Thermodynamics.

What is the mantra of pure capitalism? 'You want it? You earn it!' To make this perfectly clear: to earn means that you produce or provide value to people who are voluntarily willing to pay you. If you are greedy, you must work harder or smarter to provide additional value in order to satisfy your desires. This raises a very interesting point. There is probably little difference in the scope of one's desires independent of whether you are a collectivist or a capitalist. But there is a fundamental difference in attitude. A capitalist decides, at some point in life, that he or she is unwilling to work longer or harder in order to satisfy a desire. Doesn't mean that they wouldn't like a boat or tools or an exotic vacation, it means they have decided that the price, in terms of time and stress, is more than they are willing to pay. In other words, desires tend to be self-limiting.

In a collectivist milieu desires are not *self-limiting*. They are *rationed* by authority. A collectivist has bought, hook line and sinker, the idea that his or her needs represent a claim on the state — and hence a claim on the resources of those people called taxpayers from whom the state extracts money and time. History is clear on this: once a benefit is provided it is seldom legitimately rescinded. The benefit becomes untouchable, (social security, for example), and now the focus of collectivists is to increase payouts, develop new programs and never to look back — until the state goes bankrupt. Basically, the lesson is this: the Law of Supply and Demand guarantees failure when demand is allowed to increase unchecked. It is a collision between the finite nature of resources and the infinite realm of demand and desire.

Albert Jay Nock once made an interesting observation, still true today. It is common, he said, to read a newspaper article reporting an expensive fiasco managed by inept bureaucrats. Yet, he said, in the next column an editor might be calling for government intervention to fix a similar problem. As Einstein may have remarked, repeating a failed action with an expectation of a different result is a good definition of insanity.

Isn't it strange that we know big public projects almost always cost more than first announced, are completed later than scheduled and often don't perform as promised, and yet we don't complain as loudly as we do when a private carrier is late with a package? Isn't it obvious that society as a whole has developed a mild form of schizophrenia resulting in double standards with reference to performance by private enterprise vs. the public sector?

Galambos suggests that the widespread distrust of capitalism is partly due to the intellectual weakness of those who believe in the profit motive and who try to explain the superiority of a social structure that uses profit (rather than force or fraud) as a motivator. No insult is intended to such giants as Von Mises or Hayek. Their error, says Galambos, was to look upon profit as solely an economic objective, *which it is not*.

He defines profit as any increase in happiness acquired by moral means.

Obviously financial profit is a subset of this definition. It is not, believe it or not, the most important kind of profit. No? Then what is? What about intellectual achievements that pay increased dividends year after year with no end in sight? Where would the human race be had not a small segment of the population had the brains and the motivation to invent the wheel, clothing, the printing press, the computer, home appliances, the internet? Let's not forget advances in medical science. This reviewer would be dead twice over were it not for two medical procedures developed by professionals whose pursuit of happiness was to search for life saving solutions to fatal diseases. As a side note, physics played a major role in one treatment. Particle accelerators produced a short-lived radioactive isotope ideal for treatment of internal tumors. I have a major debt of gratitude to the physicians and physicists who, to put it inelegantly, saved my ass. And, perhaps, if you as a reader get value from this review, you might send a bit of gratitude their way as well. Did these innovators make a profit? How would you feel if your discoveries saved thousands of lives annually?

Personally, that feeling of accomplishment is a slam dunk compared to mere financial gain, no matter how large. We, as a civilization, are very lucky that innovators correctly recognize that the pursuit of knowledge is far more profitable than, say, making a killing in the options market.

The discussion above reinforces the primacy of intellectual property (primary property) over secondary property as mentioned earlier. It raises an interesting point with respect to our schizophrenic society. On one hand we have intellectuals[18] who are scrupulous in giving credit to their sources, showing proper respect for the primary property of others. They tend to defend their own intellectual property with the ferocity of a lion. Often however, they do not extend the same respect to the secondary property of others. Factually, many prominent scientists are and have been socialists. They view accumulated secondary wealth with a jaundiced eye, perhaps out of frustration because know their contributions to civilization are more important than a fortune made through arbitrage. They might also think they could put the money to better use — but that's a mistake in thinking. It's the fallacious Robin Hood syndrome: it's okay to steal for a good cause. (Of course it never is.) In summary, intellectuals are capitalists with respect to primary property and collectivists with respect to secondary property. Thus intellectuals can be called primary capital-

---

[18] Galambos notes that a true intellectual never abandons the Scientific Method. It is to his advantage to admit error. Pseudo intellectuals tend to be dogmatic, sometimes arrogant and often willing to support the party line of whatever milieu they are in. They are the arbiters of 'political correctness,'a magnificent oxymoron. Political correctness is simply a new term for that old standby; heresy.

ists who often advocate the plunder of secondary property.

The other side of the coin is a person who calls himself a capitalist and who owns or manages a business which provides a product or service desired by a segment of the public or by other businesses. The public may think his or her job is easy. It is not. An entrepreneur must juggle the competing interests of customers, employees, vendors and investors. In addition, there are tax collectors, inspectors and a plethora of interesting (and conflicting) regulations. From personal experience I can wholeheartedly say I have the greatest respect for a business person who has always met payroll. Business people tend to be very scrupulous with regard to secondary property. Cheat a customer or a vendor and shazam, the word spreads. Soon the business has problems. It is simply good business to handle secondary property honestly. Unfortunately, business people may not be as honest when it comes to primary property, namely the ideas of others. Reverse engineering, avoidance of royalty payments, patent theft and design copying are all common ploys in the world of commerce. Thus business people are partial secondary capitalists.

How Do You Know You Are Right?

No wonder we have a schizophrenic society: some of us are enamored of ideas, caring little for wealth while others seek wealth and think that the ideas from whence wealth is generated are a dime a dozen.

This dichotomy is a powerful contributor to the public's belief that there's something quite wrong with capitalism. It isn't the only factor in this propaganda mess. Let's consider 'crony capitalism.' As many have noted, there is a revolving door for influential individuals between federal bureaus and the private sector. In addition, retired legislators often become lobbyists skilled at the political representation of special interest groups. The results of these incestuous relationships seldom benefit consumers and taxpayers. Further, who gets the blame when bureaucrats or legislators pass a law to set a minimum price for, say, milk? Well, who else? It's the greedy dairy farmers and distributors who are publicly castigated because milk for babies is more expensive than it need be.[19] Somehow the coercive role of the state in enacting and enforcing such consumer unfriendly

---

[19] Industry associations may need a higher price for their products in order to pay campaign contributions.

laws is overlooked. Aren't we smart enough to assign the blame jointly and fairly?

The traditional definition of capitalism is "an economic and political system in which a country's trade and industry are controlled by private owners for profit, rather than by the state."[20] There are two defects in this definition. First, it is too narrow. It omits the ownership of ideas. Second, it implies a connection between politics and ownership which in turn implies that coercive control of property by non-owners is legitimate. But it is correct in a respect that few people seem to recognize. *Each and every one of us is the manager and owner of a one person business seeking profit by offering our personal services to others.* We can enter a contract with but one customer in which case we're an employee and we perform as the employer asks. Don't like this arrangement? Develop your skills, find a better employer or figure out a way to become independent. The strange thing is, should you become an entrepreneur rather than an employee, you will have left behind a single boss and gained a multiplicity of bosses. Your customers now tell you

---

[20] New Oxford American Dictionary.

what you need to do to be profitable — if you listen to them.

Galambos defines total capitalism as that societal structure whose mechanism is capable of protecting all forms of private property completely. This definition is given as a contrast to the current definition above. Further discussion is beyond the scope of this review. The definition is consistent with the prior definitions and postulates and may stimulate some individuals to innovate in new directions.

In the final analysis we are all capitalists. Total capitalism exists when all owners are in 100% control of their property.

Not to brag, but I myself am really quite skilled at lying and I can tell you how it's done. Like a magic trick, you distract from the sleight-of-hand by focusing attention on the irrelevant.
~ Kinsey Milhone as told to Sue Grafton

Monopoly

Let's see what the absolute standard of rightness tells us about the many concepts called 'monopoly'. What is a monopoly? The standard definition is: the exclusive possession or control of the supply or trade in a commodity or service.[21] Wikipedia then states, "Monopolies are thus characterized by a lack of economic competition to produce the good or service, a lack of viable substitute goods, and the existence of a high monopoly price well above the firm's marginal cost that leads to high monopoly profit." Like many things we were taught in kiddie school, this analysis is badly flawed.

Something has been omitted from the standard definition and analysis. Isn't it fair to note that the definition does not cover the monopoly

---

21 Apple dictionary

nations have on the use of force and fraud? Isn't it a fact that states also use their coercive power to regulate supply or trade in commodities and services?

One could go on for pages about the harm done to businesses, consumers and taxpayers by state created monopolies but that isn't the point. The point is the distinction between coercive monopolies, non-coercive monopolies and faux monopolies.

A coercive monopoly is sheltered from competition and is sustained by legal penalties of various kinds.

A non-coercive monopoly has nothing to sustain it other than the right of ownership of the product or service.

A faux monopoly is the cry of 'monopoly' used by envious people when a large company is thought to have too large a share of the market. General Motors was so labeled in the last century when they had a little over 60% of the market. When something doesn't fit the definition it is simply name calling for political purposes.

How Do You Know You Are Right?

If we go back to the first paragraph, we can readily see that making a distinction between the three kinds of monopolies will change our understanding of monopoly.

Once again, the Wikipedia statement was: "Monopolies are thus characterized by a lack of economic competition to produce the good or service, a lack of viable substitute goods, and the existence of a high monopoly price well above the firm's marginal cost that leads to high monopoly profit."

In the case of coercive monopolies, we certainly understand *why* there is a lack of economic competition; the use of force prevents it. The price to buyers may or may not be excessive, varying from nation to nation and depending on the degree of subsidization. Such monopolies can be relatively benign price-wise or viciously cruel depending on the mood of the people in power. In any event, coercive monopolies seem to be immune from criticism because the public has purchased the idea that such a monopoly is in the public interest. In the long run, however, such state run endeavors cannot endure. If they sell below cost, for example gasoline in Venezuela, the state eventually goes broke. If they sell at too high a price they have few buyers and many protesters, a circumstance unfavorable to those in power. Further, high prices always create black markets. Cigarette boats didn't get their name because they smoked, you know. Thus, state protected monopolies, while noxious, do have their limits.

Non-coercive monopolies are what most people think of when Teddy Roosevelt types attack private enterprise. Who wants to be gouged? Citizens often crave a white knight who will protect them from rapacious capitalists who wish to extract their last dollar.

This point of view is wrong because non-coercive monopolies have even more stringent limits than coercive monopolies precisely because of the profit motive. Many people are certain to ask, "Isn't it true that the higher the price, the more money a company can make?"

The answer is no. If a firm sells below cost they will eventually raise prices or go out of business. If their price is too high, sales will drop and their total profit will diminish. It will be less than the total financial profit at a lower price point. Further, marketers know that high prices attract competition which gives them an additional reason to ask a lower price. Thus the market offers a win-win solution to buyers and to sellers.

From a deeper point of view, what are some examples of exclusive possession or control of the supply or trade in a commodity or service? Perhaps the largest monopoly earners are to be found in the entertainment industry. Individual performers develop a persona that's worth millions and one which may go on earning well after the performer's death. Do we begrudge an entertainer's success? Do we say they make too much money? Don't we instinctively appreciate the preparation and the work they did to polish their performance? Why, then, do we apply a different standard to other forms of endeavor such as invention?

Once again, our society is somewhat schizophrenic. We have been told by those in power and their friends in the media that one person's need represents a claim on the property of another. Therefore there is a difference between a necessary product such as electricity and a froufrou product such as entertainment. It is claimed that one industry needs controlling for 'the public good' and the other does not.

How Do You Know You Are Right?

This point of view is not right because it validates one form of ownership while denying another. Aside from making a mockery of 'equality before the law,' it disrespects property. Yet protecting property is the primary justification for government. As I said, schizophrenic.

A bystander, not blinded by prejudice, nor warped by interests, would declare that taxes were not raised to carry on wars, but that wars were raised to carry on taxes.

~ Thomas Paine

Taxes

At the end of the next thousand years there are but two possibilities: either the concept of taxation will be viewed with the same horror as today we view cannibalism, or what's left of the human race will be cannibals.

The foregoing is an attention-getting prediction out of context. Out of what monstrous conceit can such an outrageous statement be made? How can practices as old as civilization possibly come to a peaceful and well-deserved end? How do you know this prediction is right? Truthfully, we cannot know precisely what might happen to the remnants of the human race should we fight a nuclear and/or biological war. We do know that nothing can possibly justify such a cataclysmic loss of lives and knowledge. We also know the damage tyrants can do and we know that major tyrants are born every generation or so. It is a clear statistical probability, over time, that another Hitler or would-be Caliph will attempt to conquer the world.

Please note: *you can have taxes without a war but you can't have a major war without taxes*.

Let's consult history. It is generally believed that taxes were originally collected by living groups to finance defense against other living groups. In many cases the record shows that citizens voluntarily contributed more than the amount asked for.[22]Their reasoning is clear; pay for defense and enjoy the rest of their property or — do nothing and risk losing everything.

Collecting resources for defense wasn't a bad idea. Very few individuals can guarantee their own twenty-four seven safety. The elderly, the infirm, women and children, these individuals are especially vulnerable and, to our ancestors' credit, they protected all to the best of their ability. Let's put it this way: those men who did a poor job of defending life and property often wound up a slave or dead. Either way their con-

---

[22] *For Good and Evil. The Impact of Taxes on the Course of Civilization,* Charles Adams, Madison Books, 1993

tribution to human progress was probably minimal. Those peoples able to defend themselves became our ancestors — along with some of the barbarians who seemed always at the gates.

Soon, what had been voluntary became obligatory. Obligatory — what a wealth of atrocities that word encompasses! For example, Syrian tyrants favored the beheading of delinquent taxpayers. Later, you became a slave if you could not pay your lawful taxes to the Roman State. In Russia, Czar Nicholas favored breaking tax delinquents on the wheel because it was bloodless and seemed to produce an excellent rate of compliance. Today we are far more kind and enlightened. Gone are the whips, thumbscrews and the rack. In their place: wage garnishment, asset seizure, padlocked businesses and imprisonment. The latter is particularly unappealing since the state offers no guarantee of personal safety to prisoners. Prisoners do, of course, get to watch television, but, as far as I know, it isn't obligatory.

Today Americans pay, directly and especially indirectly, a larger percentage of their earnings to our various taxing authorities than the colonists ever paid to the British monarchy. It is quite remarkable how feisty our founding fathers were about a rather small and infrequent property transfer tax as compared to our more or less placid acceptance of massive state plunder today. It's living proof, as the mobster said, that you can steal more with a briefcase than a gun. Fraud is more effective than force because it conceals from us the extent of hidden taxes. Further, those individuals who benefit from the spoils of taxation are always telling us how well they serve us and, by the way, their way is the only way to get positive public benefits such as schools, roads, police and fire protection and so on. They ought to recuse themselves.

What do you say to someone who believes this tax propaganda? You say, "Good gravy! Have you no imagination? Can you not think of alternate ways to get and pay for the things you want?"

"And, why are you paying for so many things that you don't want or need? Are you mentally handicapped?"

Tax debates: what to tax, how much to tax, who to tax and how to spend tax money — are a complete waste of time. It is akin to searching for The Philosopher's Stone. The absolute standard of rightness unambiguously says taxation is not right.

How Do You Know You Are Right?

The truth is incontrovertible.
Malice may attack it,
ignorance may deride it,
but in the end, there it is.
~ Winston Churchill

Interdisciplinary Thinking

Understanding complex phenomena such as weather patterns or social interactions has been beyond our competence until recently. What is meant by understanding? One can claim understanding when one is able to accurately predict future results from identifiable causes or actions. What if the causes for a given phenomena are interdisciplinary? This can be a problem because we have far more specialists than generalists. Specialists know more about their field than generalists, to be sure, but sometimes it is the integration of knowledge from disparate fields — the connections, as it were — which lead to massive re-evaluation and expansion of the totality of our knowledge.

By analogy, a new integration is similar to a common situation when working a challenging jigsaw puzzle. Sometimes pieces with a plausible shape and image are improperly connected. That makes it impossible to find an adjoining piece. Eventually the error is discovered, pieces are rearranged and the intended picture emerges.

Let's take an example from the real world. Thomas Gold, past president of the American Physical Society, integrated knowledge from planetary physics, biology, geology, chemistry and petroleum engineering to advance a powerful hypothesis about the origin of subsurface hydrocarbons and coal.[23] Naturally this was not greeted by cheers of admiration from all the folks who are busy teaching that oil and gas arose from prehistoric swamps and dinosaurs. The abiogenic origin of hydrocarbons flies in the face of the widely held view that 'fossil fuels' are due to run out any year now and we had best conserve, regulate, ration, control and deny their use as much as we possibly can. Basically, fossil fuel believers are defending what they learned at school.

---

[23] The Deep Hot Biosphere, Thomas Gold, Copernicus Books 2001

How Do You Know You Are Right?

If I think about it, I can compile quite a list of hypotheses taught in school that later proved false. Perhaps you can as well. As a child I read that the solar system was possibly formed from debris left behind when our sun was grazed by another star. I was told we were alone in the universe and life arose from a divine spark. So, perhaps it would be better to continually educate ourselves in the light of new knowledge rather than relying forever upon a world picture fixed in time by a graduation date.

The fossil fuel hypothesis had much to support it; initial petroleum discoveries were very shallow and fossils were found in both oil and in coal. However, this hypothesis did not explain the massive methane deposits found on the bottom of many oceans nor did it explain why established oil fields continue to produce long after their estimated holdings should have been exhausted. It did not explain why oil was found in granite with no possible origin due to biological decay. And finally, hypothetical biogenic chemical transitions were energetically unlikely.

So, what new knowledge did Gold use to form his hypothesis about earthly carbons and hydrocarbons? First, we now know that methane is ubiquitous in the solar system. It is found in planetary atmospheres, asteroids, comets, and on the surface of several moons. It would have been present when the earth was formed. Being lighter and more mobile than earth's other materials, methane slowly works its way to the crust and thence to the surface. So does helium. Helium, please note, is too light to be retained in the atmosphere by the earth's gravity. It is found almost exclusively in natural gas. This is another clue that the earth is slowly outgassing. If helium were solely a surface phenomenon on Earth it would be gone by now. Instead we have a steady supply as it works its way from the depths along with other gases. Obviously, this phenomenon was unknown to nineteenth century geologists and petroleum engineers.

Here's where it gets interesting. The world of the 1880s did not know of the ecosystem found at hot water vents deep in many oceans. Now we have proof that deep bacteria can survive at temperatures and pressures that would instantly kill surface organisms. Thermophilic bacteria are, of course, functional at very high temperatures because pressure raises the boiling point of water. These bacteria (archaea) lunch on molecular hydrogen, methane and hydrogen sulfide and use oxygen from highly oxidized iron and from sulphates to produce energy and metallic sulfides. Gold suggests that similar bacteria in the earth's crust consume methane and convert it to more complex hydrocarbons. This process is ongoing as the earth continues to outgas. Gold provides calculations that fit the facts and which point the way to further discoveries. There is much more in his book. Bravo for a magnificent intellectual integration!

The story of Louis Alvarez, et. al. is similar. When they first announced the evidence suggesting that an asteroid colliding with Earth caused massive climate change which doomed the dinosaurs, they met a storm of scientific opposition. A prominent paleontologist wrote a paper published in Nature which strongly suggested that physicists shouldn't be meddling in a discipline outside physics. Guess the distinguished paleontologist wasn't much of a generalist.

As mentioned at the start of this effort, physical scientists, heretofore, have largely avoided the social arena — other than forays to feed at the public trough. Firstly, making sense of the complexity of volitional interactions is difficult. Secondly, many innovators who have greatly improved the human condition have admitted that they don't care to interact with people all that much. They prefer the isolation that is needed for productive thought. This specialization largely prevented, up to now, any real attempt to apply the methodology of modern physics to the social sphere. Galambos mentioned that his attitude as a young scientist was the common one — a disinterest in activities designed to increase one's wealth. 'Money grubbing' is a phrase sometimes heard in university. But the loss of his father put him in charge of a small business. He learned, he said, that there was a rational beauty in business and in the insurance mechanism. Additional inputs to his integration came from the writings of Thomas Paine and Arthur Eddington. Galambos' father, Joseph B. Galambos, was his mentor for the important concept and example of integrity. Taken as a whole, the probability of a scientist encountering just exactly this set of inputs is extremely low. Something to think about if you ask, "If it's so great why hasn't anyone thought of it before?" Integrations don't come often and they don't come easy.

How Do You Know You Are Right?

New ideas need thorough vetting. One has to assess whether there is value in assuming a view of a problem that differs from accepted practice. Debate can be helpful but in large part it depends upon the integrity and aims of the participants. If facts are in dispute then differences can be resolved by observation or detective work. If, as often happens in debate, adverse facts are ignored or suppressed by one or both sides in order to bolster their agenda then debate becomes argumentation — a well known waste of time. As Mark Twain might say, take Congress as a case in point.

Wasting time is deleterious to good science. Science and invention aim to save time (and lives) in the long run. Debates and arguments based on faith rage for centuries with no resolution. They can be, in fact, a form of bloody entertainment similar to bullfighting.

Thus, if you are a scientist — a master of the scientific method — these intellectual tools will enable a deep understanding of people and the decisions they make. Further, these tools provide previously unseen solutions to some of the most intractable problems of our civilizations. Don't bother with people who use defective standards of rightness.

How Do You Know You Are Right?

Democrats are the party of government activism, the party that says the government can make you richer, smarter, taller and get the chickweed out of your lawn. Republicans are the party that says the government doesn't work, and then they get elected and prove it.
~ P. J. O'Rourke

Proprietary Interest

People who own property (their lives, ideas and chattels) are usually motivated to look after it. There are exceptions. Some people are careless about their health. Some are lazy or forgetful. This can lead to a loss of secondary property. Some people are incompetent or mentally ill. Yet we observe that a large majority of folks take routine precautions against fire, theft, flood and accident. Why work to earn something if you subsequently lose it by being careless or foolish?

Owners can be said to have a *proprietary interest* in their property. The degree of interest generally matches the value the property has for the owner and usually matches the care he or she takes of the property. Please note, the value may not be what the owner could get on the open market or at a garage sale, it may be an object of sentimental value. An old watch, an antique brooch or a bowling trophy might mean more to an individual than would a new car. In general, however, owners take decent care of their cars, homes and other big ticket items because they know, *in the long run,* maintenance is much cheaper than neglect or replacement.

Home renters, on the other hand, have a proprietary interest that differs from landlords. They want the comfortable use of the property without having to deal with the annoyances of repair, maintenance, painting, and so on. Their desire is to take the minimum care needed to get their rental deposit back. Depending upon the size of the deposit and the value the renter assigns to that money, renter and owner usually find it profitable to minimize damage to the property. But as a simple fact, the different mindset of renters often results in more property damage to rental property than is the case for owner occupied dwellings.

What happens in situations wherein there is little or no proprietary interest? Disaster is what often happens. Witness the many fiascos associated with public housing and rent control. Witness the neglect evident in abandoned military bases. Expensive military vehicles are left behind in Afghanistan because it is too expensive to ship them back to the USA. Veterans are faced with substandard care from the VA because of mismanagement, incompetence and cheating . . . . the list is long and anyone who has been in the military or worked a year in a bureaucratic setting can tell you stories that curl your hair.

Ludvig Von Mises correctly pointed out that a bureaucracy can never be similar to an efficient business which ideally would take taxes and in return deliver cost effective value to taxpayers. Why? Because a bureaucracy lacks the signal from profit (conventionally defined) which would tell whether their services were needed or desired. Bureaucrats don't know and can't *really* know whether their work is cost effective because they lack that signal.

Yet, millions of federal and state employees show up on time, do their job and follow the byzantine rules of the system to the best of their ability. There *is* a motivation to serve the public — with the possible exception of the IRS. Bureaucrats, by and large, are not evil people bent on exploiting the taxpayer. Their proprietary interest stems from doing their job to the best of their ability. Why, then, has a country that was once the world's beacon of opportunity morphed into a welfare state that punishes productivity and initiative while rewarding indolence and misbehavior? It is because the top level bureaucrats *do* seek a form of profit **as they see it**.[24] They call it winning. Being elected increases their happiness and boosts their self esteem. The happiness they seek is bolstered by votes and the reward is control of the levers of political power. Decisions made when serving the public are actually finely calculated and measured as to which approach or giveaway shall result in the highest return, that is, the greatest number of votes. This is the status quo and it ain't news.

"Whoa," you may be thinking, "how can you use the term profit with regard to bureaucrats when Galambos clearly tied profit to morality and taxes are immoral?" The answer is that bureaucrats are actually seeking immoral profit not moral profit. Since they are taking and spending other peoples' money they are actually

---

[24] Profit is an increase in happiness achieved by moral means.

seeking their own happiness through the acquisition and distribution of *plunder* — a good name for an increase in happiness immorally achieved.

It is this distinction between a moral increase in happiness and an immoral increase in happiness that has given us our schizophrenic society! We have, as a civilization, allowed ourselves to confuse legality with morality. If it's legal, it must be right, right? Wrong! Some laws are certainly moral and many are not.

One of the tenets of Western Civilization is the idea that the 'rule of law is superior to the rule of man.' What we forget is that the rule of law is only one step removed from the rule of man. What one legislature can decree a later legislature can amend or expand. Thus laws proliferate and grow. They are rarely repealed. When the burden of laws and their associated administrative costs (including the mandated unpaid labor performed by shop keepers and employers) reaches a tipping point, the nation dies and the survivors start over. Well, it *is* clear that the rule of law is much superior to anarchy. And *it is also* clear that civic squabbles have more to do with who gets to make the laws than they do with what, exactly, the laws are. I don't remember that point of view being presented in my high school civics class, do you?

In summary, it is observable that the lack of true proprietary interest by people charged with stewardship of public resources leads to less than optimum management. While it is true that owners can and sometimes do mismanage their property, it is also true that they are generally better managers because they have to bear any losses due to misfortune or mismanagement. Stewards of public property do not bear losses, taxpayers do. It is precisely this lack of positive and negative responsibilities for state employees which explains the manifold failures of state run enterprises.

Here's your homework assignment: figure out how we can have schools, roads and property protection without using force or fraud to pay the bills.

Most of the presidential candidates' economic packages involve 'tax breaks,' which is when the government, amid great fanfare, generously decides not to take quite so much of your income.In other words, these candidates are trying to buy your votes with your own money.
~ Dave Barry

Sheer Speculation

Shorn of trivial items such as sports and entertainment, the most basic needs of an individual include food and water, the necessities for survival. Depending on climate, shelter is often essential to survival. Our antecedents have, aside from war, solved the basic survival problems facing the species. Today, most of us always have food to eat, something to drink and a place to sleep. We have but to decide whether to go out for a meal or to prepare one.

It wasn't always thus. Natural phenomena could cause crop or game failure followed by starvation and death. Disease was not at all understood and large percentages of the population were killed by a very long list of ugly diseases such as malaria, the black death, pneumonia, dysentery and cholera.

Animals face many of the same survival problems as do humans. Paleontologists tell us that more species have gone extinct than live today. Why did they fail to survive whilst humans have prospered? The answer is obvious; our ancestors somehow developed the ability to think in the abstract, to communicate and to preserve their thoughts. Intelligence promoted the use of tools and fire. Our ancestors found a way to survive in an environment that might kill many modern men and women. And they did it without fang and claw! Bravo!

On the other hand, civilization has created new hazards to species survival and magnified a few of the older hazards. For example, experiments with small animals show that the stress of overcrowding can lead to deviant and harmful behavior. While it is not always wise to extrapolate from animal experiments, similarities in behavior such as infanticide as a function of population density do strike a warning note.

Technology is sometimes cited as a threat to public health and welfare (think auto accidents). Technology isn't a threat. It doesn't eliminate accidents, it usually reduces them. Comparison of deaths per mile for auto travel compared to deaths per mile for horse and buggy travel is a good example of this. Not only are you safer in an auto than a buggy, but your safety improves year by year due to seat belts, air bags and collision avoidance systems. Airplane travel, in terms of fatalities per mile, is safer yet. There is one caveat: when a new technology is introduced, unforeseen hazards can cause harm until they are identified.

That said, most hazards of civilization are trivial when compared to the death and destruction associated with war, civil war and civil unrest. It is heartbreaking that millions upon millions of innocent men, women and children have been legally slaughtered by state employees as sacrifices to the ambition of psychopaths. Psychopaths clearly value political dominance far more than any feelings of remorse that they should feel — were they normal.

There is a solution to the problems of war and genocide. It isn't pacifism nor love thy neighbor. Haven't these been tried more than once? One must first understand the defects in our societal structures (and perhaps in our individual psychologies) which allow a criminal to take the helm of a ship of state. And, please note, the engines of these ships are fueled by taxes. Unfortunately, the ship's crew goes along to get along. Major redesign is clearly necessary. Can we invent voluntary mechanisms which simultaneously protect property and which cannot be used by sociopaths and psychopaths? If not, why not?

A cautionary note: some people, upon realizing that taxation is theft, go off the deep end as rebels. They may cease paying taxes and urge others to do the same. **This is an extremely bad idea.** The only way to convert taxes (and hence wars) into history book footnotes is to create superior alternatives and mechanisms for the protection of all forms of property. Further, one does not utilize coercion in any form in order to end coercion. The means must be consistent with the ends. Self defense, however, is rational and moral.

What about future hazards to the species? The reviewer is pleased to mention that he correctly identified asteroid collisions as a real time risk factor in the seventies. The existence of huge crater remnants such as the 300 km diameter Vredefort crater in the Free State

How Do You Know You Are Right?

Province of South Africa tells us that life on earth could change as catastrophically in the future as it has in the past. Note that the Chicxulub crater, which is thought to be responsible for the extinction of dinosaurs, is only 180 km in diameter. Fortunately, we are now observing and cataloging these rogue rocks. Perhaps one day we'll have the ability to safely modify their orbits. Meanwhile, humans will simply have to hope the species survives by hiding in nooks and crannies like cockroaches.

Another hyper-speculative risk factor might be the existence of intelligent extraterrestrial life hostile to humankind. The Hollywood versions of fiends from space can be awesomely intimidating. It's a wonder humans can survive or prevail against a species with the imagined ability to travel faster than the speed of light in space. Yet, if history sometimes repeats, the lesson is; expect the unexpected. Let's put this particular threat to rest. If an alien civilization had the knowledge, the energy, and the motivation to travel light years of distance to visit Earth then that civilization would have long ago discovered how to live without committing species suicide. They would have discovered, as did Galambos, that respect for property is the golden recipe for peace. Hence they would find it more profitable to interact with us than to attack us. Of course, were they to watch our television from space, they might choose not to interact with us at all. Lastly, space traveling aliens might have life spans many times longer than ours. Would they risk their long and productive lives for a contact with dangerous savages?

If you've read this far, thank you for letting me have some fun after a very serious discussion.

Basic Volitional Definitions

This is not a dictionary. A dictionary chronicles the current use and misuse of words as determined by a committee. This list departs from a dictionary in two particulars. First, it is designed to avoid circular definitions by basing subsequent definitions on a few operational definitions. These definitions, in the main, are those used by Andrew J. Galambos.

Second, only one definition is given per word. This may seem restrictive but it is consistent with the rigor used in computer programming, mathematics and physics. English is a very useful and sometimes beautiful language but multiple meanings per word are better suited to poets and punsters than to scientists.

Absolute: independent of arbitrary standards of measurement (in physics).

Absolute: independent of arbitrary standards of determination (in volition).

Absolute good: a subjective good to at least one person, which is not imposed on another volitional being.

Accuracy: the degree to which the result of a measurement, calculation, or specification conforms to the correct value or to a standard.

Authority: moral control of property either by owner or owner's representative.

Bad: an individual's subjective evaluation of a (volitional) dispreference.

Bureaucracy: any legal organization that controls the property of individuals without their consent.

Capital: property beyond the minimum for survival which is available to generate additional property.

Coercion: attempted, intentional interference with the property of another.

Contract: a voluntary agreement between two or more people who have property which they are desirous of exchanging the use and control thereof.

Collectivism: the practice or principle of giving a group priority over each individual in it. (Oxford English Dictionary)

Control: the ability to make volitional decisions concerning the disposition of property.

Corroboration: repeated agreement of experiment with hypothesis with no exceptions.

Crime: an act of successful coercion.

Epistemology: the theory of knowledge, especially with regard to its methods, validity, and scope.

Force: physical harm or the threat thereof including imprisonment. (Anon.)

Fraud: intellectual deception on any level. (Anon.)

Good: an individual's subjective evaluation of a (volitional) preference.

Govern: to exercise specified moral control over action.

Government: any non-coercive mechanism designed to protect property to which the owner of the property may voluntarily subscribe.

Happiness: the sum of an individual's goods minus the sum of their bads to date.

Hypothesis: a guess proposed as a theoretical explanation for the occurrence of some specified group of phenomena.

Idea: an articulated thought.

Importance, absolute: the measure of the total amount of property involved.

Individualism: a social theory favoring freedom of action for individuals over collective or state control. (Apple dictionary)

Law of Nature: an honor bestowed on a theory so multiply proven over time that it is no longer a topic of serious debate.

Moral: coercion is absent.

Morality: the subject that deals with the totality of absolute goods.

Observation: direct input through the five senses, augmented by technology, of the world of which we are a part.

Operational definition: one which describes the mode of determination or measurement of whatever is being defined.

Physics: the branch of science concerned with the nature and properties of matter and energy. Anon.

Plunder: any increase in happiness achieved by immoral means.

How Do You Know You Are Right?

Postulate: a proposition with a truth content which was not derived from any earlier set of propositions by logical reasoning, but whose validity is accepted as an input into the subject you are developing.

Precision: refers to razor sharp criteria which always generate a yes or no answer.

Profit: any increase in happiness acquired by moral means.

Property: individual man's life and all non-procreative derivatives of his life.

Proprietary: of or relating to an owner or ownership.

Rationality: the use of the Scientific Method in determining the criterion of rightness.

Right (physics): simultaneously true and valid.

Right (volition): simultaneously true, valid and moral.

Rule: immoral control over property by edict.

Slavery: a condition in which an individual loses control of any part of their property through coercion. The loss need not be total. For example, taxation or conscription seldom takes all of one's property or time.

States: social mechanisms which coercively extract property from citizens for the ostensible purpose of protecting the remainder. (Albert J. Nock)

Theory: a multiply corroborated hypothesis.

Total capitalism: that societal structure whose mechanism is capable of protecting all forms of private property completely.

True: that which is observable.

Valid: conforming to the rules of logic.

Volition: the act of choosing.

Volitional Science: a substitute name for social science. Just as the name astronomy was coined to differentiate the scientific study of the stars from the pseudo study called astrology, a new name for the scientific study of volitional inter-actions was needed to differentiate this science from prior trial and error analyses and proce-dures.

Supplemental Reading List

Sic Itur ad Astra, Andrew J. Galambos, The Universal Scientific Publications Co., 1998

The Philosophy of Physical Science, Arthur Eddington, Cambridge University Press, 1939

Thrust for Freedom, Andrew J. Galambos, The Universal Scientific Publications Co. 1999

Economics in One Lesson, Henry Hazlitt, Harper & Brothers Publishing, 1946

Bureaucracy, Ludwig Von Mises, Arlington House, 1969, also Yale University Press 1944

For Good and Evil. The Impact of Taxes on the Course of Civilization by Charles Adams, Madison Books, 1993

Additional sources:

The Free Enterprise Institute, www.fel-ajg.-com

How Do You Know You Are Right?

Quotes are from the Kindle edition of *Nifty Quotes for Contrarian Folks.*

An innovator finds it harder to understand,
as a rule, why people do not understand
a discovery than it was to make the discovery.
~ Hermann Von Helmholtz

Afterword

Longer ago than I'd like to recall, William Wheeler, my internationally known boss at Varian Associates in Palo Alto, asked me if I'd like to attend a lecture at his house in Saratoga. California, at that time, was a hotbed of touchy-feelie seminars led by silk smooth entrepreneurs catering to directionless individuals. Naturally I was instantly on guard even though I had the utmost respect for Bill's competence.

"What's it about?" I asked.

"Well," he said, "that's hard to answer. It's about parents and children, societal diseases such as war . . . It's an explanation as to why so many well-intentioned ideas go so dreadfully wrong in today's societies." He went on in this rather vague way for a bit and concluded by saying that much of what we knew today was wrong.

How Do You Know You Are Right?

Well, I'm always game to learn something new — and there was a science fictionish hint of the unknown in what he was trying to describe. So, out of curiosity and also respect for the man, my wife and I agreed to attend a three lecture weekend for $15.00 each. That was about the cost of an office visit to a doctor at the time.

The lecturer turned out to be a young, balding, genial, six foot six fellow who gained our attention by a series of outrageous claims as well as a listing of incontrovertible facts. My notes, I'm sorry to say, were not the crisp, concise notes taken in thermodynamics or relativistic physics, they were four pages of skeptical rebuttals . . . plus a plentiful supply of doodles.

"So," said my wife at the end of the day, "shall we come back tomorrow or let it go?"

"Let's come back," I said. "He spent quite a bit of time on the scientific method. I'd like to see where he's going with this." My curiosity had been further stimulated and — having more than one ancestor from Scotland — I certainly wanted to get my money's worth.

Parenthetically, I recall asking the lecturer, Jay S. Snelson, what his background might be. Before lecturing for the Free Enterprise Institute he had been a photographer. Yet he was presenting a clearer picture of the foundations of modern physics than I had ever heard at university. Puzzling, a photographer who does a better job with the epistemological foundation of physics than my professors?

The second day brought the disclosure of the Absolute Standard of Rightness. I was hooked. I knew, from courses in philosophy and physics, this was a major discovery. Why? Because philosophers prefer to pontificate rather than to test their ideas in the real world. In addition, their definitions are often unrelated to the real world, hence of uncertain value. Physicists, on the other hand, want little or nothing to do with human volition. Give them a something reproducible to measure and, with luck, the boundaries of human knowledge will be ever so slightly expanded. And sometimes knowledge leads to genuine revolution. Example: discoveries in solid state physics have revolutionized computing and communications.

Here's an important point. Almost without exception scientists tend to compartmentalize. Not only do they not know much about an unrelated field, they may not really want to know. It is extraneous and thus a distraction. Hence one must be a generalist to discover or to appreciate a cross-discipline advance in scientific understanding.

That said, this might be a place to mention some of Galambos' history relevant to his discoveries. His father, Joseph B. Galambos, was a Hungarian architect who served in World War I. In the thirties, Mr. Galambos correctly predicted the advent of World War II. He was able to move his wife and young son to New York City in time to avoid that tragedy. Ironically, when WWII did break out, Andrew was old enough and patriotic enough to join the army. After the war, he earned a graduate degree in physics but family responsibilities after the death of his father eventually took him to a job in a Southern California aerospace company. His primary interest was space exploration but, as so often happens in science, fate took a hand as he was teaching evening courses at Whittier College. His strong views on societal flaws led to a lecture series at Whittier and later to the formation of the Free Enterprise Institute. Using the absolute standard of rightness, he has revolutionized social science and developed non-coercive mechanisms for the protection of all forms of property, especially ideas. Eventually his work will produce a renaissance of rationality. In any event he will be recognized as one of the twentieth century giants of science.

How Do You Know You Are Right?

Acknowledgements

The most important contributor to this effort is, of course, Andrew J. Galambos. Less obvious contributors are the giants of the scientific past such as Newton, Bruno, Archimedes, Galileo, Einstein, Darwin, Watson & Crick . . . plus the totality of innovators and inventors upon whose shoulders we stand.

Thanks to William Wheeler who introduced me to Jay Stuart Snelson who ably taught Galambos' discoveries.

Willis Lamb, Felix Bloch and my University advisor Robert Hofstadter provided background which enabled recognition of Galambos' work.

Richard Feynman and Gale Snouse provided motivation for the writer to pursue this very difficult task. James Martin drew my attention to the Garfield incident. Donald Beck made helpful suggestions designed to improve comprehension. And finally, Cathy Pettijohn Crook who did a wonderful job of editing — and who proved to the reviewer that he knew far less

about grammar and punctuation than he
thought.

T. Snouse

*It is as fatal as it is cowardly to blink (at) facts
because they are not to our taste.*
~ John Tyndall

Addendum - The Scientific Method

This review is aimed at professionals in physics, engineering and social science. It is assumed they are well acquainted with the method. For the benefit of the casual reader, a recapitulation of an explicit form of the Scientific Method is listed here.

1.   First we observe for the purpose of data gathering. We observe using our five senses augmented by technology such as a microscope. We gather, organize and test the facts much like a private detective assembling clues.

2.   Now we ask if there is a connecting principle which illuminates and explains the data. We formulate a hypothesis using valid thought processes.

3.   We then extrapolate to predict future behavior or measurements.

4.   Finally, we observe again and again to compare predictions with reality. If we have agreement again and again with no failures, we may then compute the degree of confidence based on the number of successful observations. If there is as little as one failure we have a faulty hypothesis.

Galambos makes the point that the first and fourth steps are observational and the second and third are intellectual. The first and fourth deal with truth, the second and third deal with validity. Today we have experimentalists working the first and fourth steps while theoreticians consider the second and third steps to be their territory. He dislikes this separation and suggests that compartmentalization hinders the development of new knowledge.

A similar situation exists in modern medicine. Specialists become accustomed to seeing medical customers with symptoms thought to match their specialty. When something unusual comes along, such as scurvy or leprosy, it is very likely to be misdiagnosed.

About the Reviewer

Thomas Snouse is a retired physicist, product developer and entrepreneur. He spent ten years at NASA Ames doing basic research in ultra-high vacuum and ion bombardment of metals. After a brief hiatus in Hawaii collecting reef fish, he became part of a team at Varian associates which developed award winning, commercially successful high vacuum pumps. He was a senior member of the American Vacuum Society and a member of the American Physical Society.

Currently living in Lompoc California, he divides his time between writing and playing tournament bridge.

How Do You Know You Are Right?

Coming Soon:

Common Sense and Courage